STEAM GOLD

GWR Castle-class locomotive No. 5043 *Earl of Mount Edgcumbe* arrives at Birmingham Moor Street station after a couple of runs to Stratford-upon-Avon and back.

STEAM GOLD

A NEW AGE FOR PRESERVED STEAM

GEOFF SWAINE

The History Press

First published 2017

The History Press
The Mill, Brimscombe Port
Stroud, Gloucestershire, GL5 2QG
www.thehistorypress.co.uk

British Library Cataloguing in Publication Data.
A catalogue record for this book is available from the British Library.

ISBN 978 0 7509 8240 5

Typesetting and origination by The History Press
Printed and bound in India

CONTENTS

Cover illustrations
Front: A3 4-6-2 (Pacific) No. 60103 *Flying Scotsman* flies along the East Coast main line. *Back*: WD 90733 2-8-0, an engine thought to be so mundane that none were saved. However, some very keen enthusiasts found this one in Sweden.

Opposite: Black 5 No. 45231 *The Sherwood Forester* runs into service at Minehead.

K1 class No. 62005 crosses the famous Glenfinnan viaduct on the West Highland line, Scotland.

INTRODUCTION

It is hard to believe that now, almost fifty years on from the last passenger steam service in the UK, we find ourselves in a new golden age for steam railways. Not since the 'Fifteen Guinea Special' pulled back into Liverpool Lime Street in 1968 has the sight and sound of these marvellous engines been so accessible and so widely appreciated.

Preservationists grasped the severity of the situation early – back in the 1960s when steam engines looked as though they were going to be lost forever. Judging from attendances at new heritage centres, these machines continue to hold a special place in the minds of the population.

What makes us love these old railways so much? Well, it surely has something to do with the social and physical history of these small islands. Our railways emerged through private companies that looked for profit and were keen to exploit the situation to the full. They also needed to offer something different to their rivals, but the new industry was so intense that they had no qualms about duplicating services. Railway Mania (the fast expansion of rail companies) came about as the general public quickly realised that railways could transform their lives. No longer were people confined to their local surroundings; now they could go to the coast, or to London – as so many did in 1851 to see the Great Exhibition. Even if they didn't travel themselves, most people benefited from the advent of this transport system: the supply of fresh foods became available

for the first time in many areas; fish could be landed at ports and arrive in markets the next day; vegetables grown in distant parts of the country were brought into towns and cities; and cattle could be transported, rather than be herded 'on the hoof' for days. This, together with being able to travel to the newly developing seaside towns, really did inspire people. They saw it, liked it and used it. Life was never to be the same again.

As well as the public, governments loved the railways because of the boost they provided to industry; the 1820s were tough times, but new railway building plans got things moving again.

With so many private companies competing against each other, none were content just to be the same as their rivals; they wanted to be better. Lines competed against each other to cover the same routes and to serve the same towns, each insisting on having its own station. It was usually down to the boss of each company as to which building style it would use to construct its stations. We can still see the effect of this today, with plenty of stylish stations still standing, built in architectural styles ranging from Gothic, to Roman and Greek classical, Georgian and on towards the modern. Modernists, rejecting the more ornate traditions of Gothic architecture, preferred to design in a simple manner, with the exterior of the building showing its function. King's Cross station, for example, which sits alongside the outright Gothic St Pancras, clearly shows this trend. Many company chiefs still opted for Gothic, however, as they considered it the best means of showcasing their railway. The front elevation of a major station building was its architectural showpiece, much like the west face of a cathedral, and this individualism ran right through the railways' history until the post-war nationalisation of the railways in 1948, when standardisation became the norm. Up to this point, the impetus was towards variety; the design of engines were differentiated, with every company developing its own style, all wanting to be just that bit different – even the sound of the whistles was different.

Now, thanks to the preservation movement, we have the chance to really appreciate this rich history, with many machines having been restored to their glorious best. It is so different from those post-war years when the railways lost out to the arrival of the family car, leading them to fall in to decay, becoming neglected and covered in grime. The steam engines had to go, but we must be grateful that there was a swansong through the 1950s and early '60s, otherwise we wouldn't have what we do today. After the war, many European countries rebuilt their systems using modern technology, utilising diesel or electric locomotives as the driving force. Their systems – mostly bombed out and derelict – required this sort of attention, but Britain's railways had kept going all through, and long past, the war years, despite the fact that they were mostly run down and worn out. Foreign oil was unaffordable, so the reliance on home-grown coal continued. As a result, the British railways had a reprieve, which lasted another twenty years.

It is clear from the many popular re-enactment events held across the country that our railways, and the history surrounding them, have captured the imagination of the British public. Without doubt, we now find ourselves in what must surely be the last golden age of steam – a late flourishing of machines and appreciation – of which this book is a small but enthusiastic testament.

PRESERVATIONISTS TRIUMPH

A3 4-6-2 (Pacific) No. 60103 *Flying Scotsman* passes Conington on the inaugural re-launch after renovation along the East Coast main line, 25 February 2016.

Flying Scotsman passes Conington level crossing, heading for Peterborough.

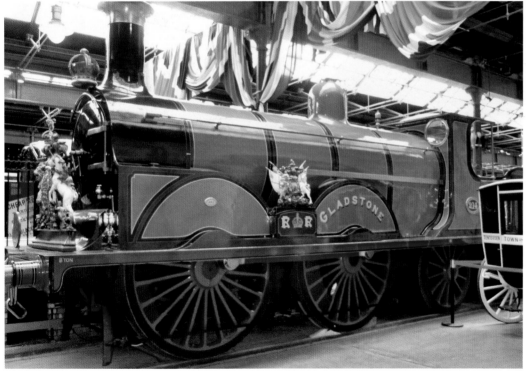

London Brighton & South Coast Railway Stoudley B class 0-4-2 No. 214 *Gladstone*, at the National Railway Museum, York, in supreme gold-ochre livery with royal insignia. This livery was designed to mark seventy years of preservation.

EARLY PROGRESS

The original golden age of the railways spanned from the time Railway Mania started in the mid 1800s through to the Edwardian period, when it was dramatically curtailed due to the outbreak of the First World War. It was the Edwardian era that really encapsulated just how the nation had taken to its railways – a period of evolution after many early lessons had been learned from the Victorian era before it.

The earliest steam locomotives derived from Cornwall in the south-west peninsula of England, an area rich in the mining of tin and copper, these products having been mined here for centuries past. In the 1700s static steam engines had been developed that pumped water out of the deep mines, as the area was near the coast with water being the obvious problem, and much of the mining was below the sea. With the introduction of steam-powered pumps, mining companies were able to exploit the ground to much greater depths. This was a great thing for the mine owners, whose bottom line was always about making money – for themselves.

It was a local man, Richard Trevithick, who first had the notion that a steam engine could power a moving vehicle. To do this, he realised, the steam engine would have to be powered by high-pressure steam – all the static engines operated at a very low pressure, and never more than 4lb psi. Trevithick knew that when pressure built up in a boiler steam became a very dangerous substance, and increasing pressure within a boiler would eventually lead to an explosion, unless, that is, the boiler had a suitable safety valve to release the excess.

Right: The royal coat of arms is part of the royal insignia added to this train.

Far right: Thompson LNER B1 class No. 1306 (BR 61306) *Mayflower* (1948) brings home a service into Shackerstone station on the Battlefield line near Leicester.

Left: After a recent overhaul and a new ten-year running ticket, 'Modified Hall' No. 6990 4-6-0 *Witherslack Hall* runs into Leicester North station, the southern terminus of the Great Central Railway. Note the Hawksworth post-war flush-sided tender.

Right: Pulling off her train for the return journey to Loughborough. No. 6990, 4-6-0 *Witherslack Hall* prepares to run around the train. The 'Halls' were the mid-range workhorses of the GWR. The modified type were the last range with improved superheating. At the front, the cross plate between the bogie wheels and the triangular stays above the buffer beam are the visible differences to those produced previously.

Static steam engines had been developed to haul wagons along wooden tracks by means of a cable, especially in the north-east of England where coal mining was a big industry. Trevithick tried to produce a road-running vehicle powered by steam, but it was ungainly and quite unsatisfactory. His first successful moving steam engine had a boiler pressure of 40lb psi.

Trevithick did produce his locomotive to run on rails, which was exhibited in front of a sceptical public in 1802 – how could an engine with shiny metal wheels run on shiny metal rails? Indeed, it was the rails that were the major problem. Cast iron was the only product available, which, as everyone knew, was brittle and broke easily. It was brilliant for station canopy structures etc., high up where the

moulded product could stay for centuries unstressed; it was brilliant in compression but poor in tension. However, down on the ground, with heavy trains pounding across, it was completely unsuitable. Cast iron had taken over from wooden wagon-ways, which were suited to horse-drawn trains where the stresses were light. The fish-bellied shape helped, but for a heavy locomotive full of water there was no chance.

Wrought iron was a product well known by blacksmiths as a workable product, where the impurities of cast iron are removed in the smelting stage; the product becomes strong in tension and malleable. In 1820 John Birkenshaw developed a rolling mill capable of producing long, thin sections of the product. Development was

rapid with wrought-iron rails until steel became available from the 1850s onwards.

By the 1840s Railway Mania had set in, producing a rail network capable of feeding the entire country. The railways provided widespread mobility as well as making perishable food items freely available for the first time. Fish and chip shops emerged in the thousands to offer the public a meal, which has since stayed the nation's favourite for over a century. The steam engine found its way into other areas such as entertainment, providing steam-driven rides. Farming benefitted from the development of the traction engine, the first piece of mechanised equipment to aid the farmer and lessen his reliance on the horse.

By the second half of the nineteenth century, style had crept in to the design of railway rolling stock. Main stations became the new palaces and cathedrals of the Empire. Seaside towns changed from little villages to resorts, and company owners were eager to cater to the public's burgeoning appetite to travel.

By the Edwardian period, so popular and well used was the transport system that the private companies competed to add luxury, style and elaborate liveries. Engines were loved and cherished, so different from what would happen fifty years later when, nearing the end of their life, all the locomotives and stock went into decline.

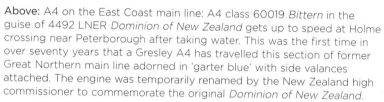

Above: A4 on the East Coast main line: A4 class 60019 *Bittern* in the guise of 4492 LNER *Dominion of New Zealand* gets up to speed at Holme crossing near Peterborough after taking water. This was the first time in over seventy years that a Gresley A4 has travelled this section of former Great Northern main line adorned in 'garter blue' with side valances attached. The engine was temporarily renamed by the New Zealand high commissioner to commemorate the original *Dominion of New Zealand*.

Above right: Borrowing an identity from No. 60019, *Dominion of New Zealand* looks the part on the East Coast main line.

Right: No. 60007 LNER A4 class 4-6-2 *Sir Nigel Gresley* occasionally tours the country and is shown here in Scotland. The train is marshalled as a push–pull unit for the journey between Boness and Manuel. Here the unit is moving out from Boness with *Sir Nigel* taking up the rear.

Above and below: Classic country trainspotting, with 'West Country' Pacific No. 34007 *Wadebridge* taking an eastbound service along the Mid-Hants Railway. The train passes before the silence is broken by Schools-class *Cheltenham* coming the other way. Resident of the Mid-Hants railway, 'West Country' Pacific No. 34007 is heading for Alton where there is a main line connection. This example is still very much in its original form; the class was designed by the controversial Oliver Bulleid.

Above and below: The three-cylindered Schools class No. 925 were a highly successful class of engine named after public schools. Forty were built in total, with three surviving into preservation. The class was built by Maunsell, from 1930 onward, with the width restrictions on the London–Hastings line in mind. Maunsell took a chance on the 4-4-0 wheel formation but in the end they outshone even the Lord Nelsons. No. 30925 *Cheltenham* is passing Bowers Green Bridge heading westward.

Left: Tangmere on the West London line: unrebuilt Bulleid 'Battle of Britain' Pacific 4-6-2 No. 34067 *Tangmere* hurries along the West London line on a sunny spring morning. SR No. 34064 (built in 1947 in Brighton) is here heading a 'Golden Arrow Statesman' Pullman special to Canterbury and back. The classic engine in full regalia makes a fine picture passing the Olympia exhibition centre.

Below: Rebuilt Bulleid 'West Country' Pacific 4-6-2 No. 34046 *Braunton* arrives at Kensington Olympia during a rail tour. With the Grand Hall of Olympia in the background No. 34046 waits just long enough to pick up some passengers before moving off with the steam-tour into Kent. This section of line is one of the main north–south connecting routes between regions and was a key goods and military artery during the war when it was closed to passenger services.

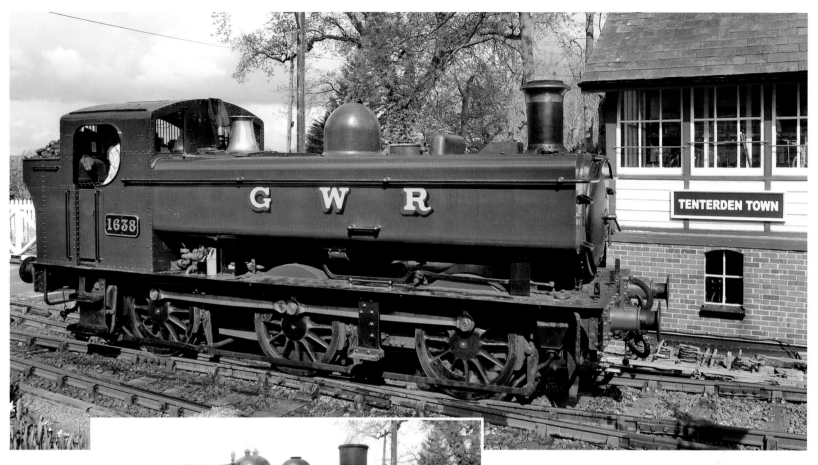

Above: Resident at the Kent & East Sussex Railway, GWR Pannier tank No. 1638 passes the signal box at Tenterden – a late Hawksworth Pannier built in 1951.

Left: Switching tracks at Tenterden is SR USA class No. 65. This shot gives a good view of the engine's very short wheelbase, which meant that it was well suited for working the sharp curves around the former Southampton Dock lines.

LAST DECADES

Owing to the state of the railways after the war, it is not surprising that they became public property with nationalisation. The owners of the 'Big Four' railway companies – London, Midland and Scottish Railway (LMS), London and North Eastern Railway (LNER), Great Western Railway (GWR) and Southern Railway (SR) – who had briefly had their power reinstated after the government control of wartime, had no hesitation in taking a pay off and getting out. Instead of there being four private companies operating in regions, the railways would continue in the same manner but with the territories now redrawn as Eastern Region, Southern Region, Western Region and Midland Region, all of which were placed under public ownership. Eventually there would also be a North Eastern and Scottish Region.

Robert Riddles, who had produced the WD types during the war, was brought in to be, in all but name, the chief mechanical engineer for British Railways. He was charged with formulating designs of engines needed to produce examples for all power classifications. The power classifications started with the lowest being '1', and going through to the most powerful, which would be '9'.

A Conservative government may have taken a different line, but a large proportion of railway people wanted nationalisation. To the GWR this meant the end of a company that had operated with pride for over 100 years. Its staff didn't see it coming, and the public ownership spelled the end of Swindon as a prime locomotive-producing town; however, much of Swindon wanted, and so welcomed, public ownership.

One of the most enduring locos ever: the Stroudly Terrier No. 3 *Bodiam* at Tenterden. The class originally hauled early commuter trains around south and east London. Many survive to this day, proving how successful they were.

The Tyseley Loco Works hold annual open days where they not only show off their special fleet but also bring in some guest exhibits from other regions. Here the master of ceremonies stands proud whilst controlling all the movements on and off the turntable. Castle-class locomotive No. 5043 *Earl of Mount Edgcumbe* is on the turntable.

No. 4965 *Rood Ashton Hall* was once thought to be *Albert Hall* before some identification on the frame proved otherwise.

On the demonstration line at Tyseley is the main line favourite, 5700 class Pannier No. 9600. The engine carries the black BR livery of an express passenger train.

The production of the British Standard classes didn't start until the early 1950s, so much activity was needed before that. Chapter 5 in this book gives more details of the locomotives that came through the austerity period after wartime.

Robert Riddles, an ex-LMS man, had worked under LMS Chief Mechanical Engineer William Stanier before the war and so the influence from that railway company was imported into his designs; even the power classifications were ex-LMS. He set out to produce twelve classes of locomotives to cover the whole spectrum of necessary requirements. The early showstoppers were the Britannia class, one of which, No. 70034 *William Shakespeare,* appeared at the 1951 Festival of Britain exhibition on London's South Bank, a showpiece exhibition where the Labour government set out to show off a new modern future for Britain. No. 70013 4-6-2 *Oliver Cromwell* has survived into preservation, mainly because it was active right up to the end of steam in 1968. Also, the prototype 70000 *Britannia* is still with us and is running once again on the lines.

In total, 999 examples of engines were produced. The railway public didn't immediately take to them, partly because it was obvious that the steam engine had a limited lifespan in a fast-changing future. In other countries, impressive electric and diesel traction engines were being installed, especially when building from scratch after the hostilities.

The most numerous class of British Standard engines was the 9F. The 2-10-0 heavy-goods engines also proved excellent performers on passenger train duties. Hence, with their newness, many of the worn-out older engines could head for the scrapyard. In all, 251 9Fs were built. Nine survived into preservation, including the last to be built by British Railways, No. 92220 *Evening Star*, which was out-shopped from Swindon in 1960. These last engines had a very short life, owing to the government's change of heart towards steam – a change brought about by the mounting losses in operating the system. A radical change was enacted, coming to light with the issuing of a 'Modernisation Plan'. The White Paper of 1956 stated (amongst other things) that all steam engines should be phased out from Britain's railways as soon as possible. In fact, the last steam-hauled service was in 1968, which shows how fast things were to happen – especially as steam engines continued to be

built until 1960. The lack of appreciation of how road traffic was to further lessen the need for rail travel was also badly underestimated, whereby massive investment was put into the construction of large marshalling yards for goods traffic – this was when freight traffic was increasingly coming onto the roads. These last locomotives coming out of the rail workshops were capable of having an economic life of thirty to forty years, but in reality most had a life of less than eight.

The swiftness in all this led to an uprising of discontent from a certain age group of people. Young people who had admired steam so enthusiastically through the last decade, even though everything was grimy and in decline, stood together in their discontent. Beeching was to come, and so were the new preservationists.

The run between Moorgate and Hammersmith is the favourite for bringing steam back into the tunnels of London. Waiting to take a train away from Hammersmith for a return run is Metropolitan Railway E class 0-4-4T No. 1. Note the carriage behind the engine is Metropolitan milk van No. 3 (for taking churns), which dates from 1896.

London Transport-liveried former GWR small Prairie 2-6-2T No. 5521 (with a reduced cab), today running as L150, cruises into Rickmansworth station. The former GWR engine was called in for the Chesham branch line to be the second engine on a special two-day schedule. Behind the Prairie is electric loco *Sarah Siddons*.

GWR small prairie is now at the rear of the train as it departs Chesham with a return trip to Rickmansworth. No. 1, which is at the head of this train, worked the last steam train on the Chesham branch in 1962. The steam workings of this weekend in August 2014 were the first since that date.

Small Prairie GWR No. 5552 sits just outside Bodmin General station on the Bodmin and Wenford Railway. The full height of the cab can be compared here.

Preparing for a Keighley & Worth Valley 1940s weekend. In the foreground is WD 90733 2-8-0, an engine thought to be so mundane that none were saved. However, some very keen enthusiasts found this one in Sweden.

Two marvelous war engines sit together at Haworth: War Department No. 90733 and USA Transportation Corps 'S160' 2-8-0 No. 5820 Big Jim. Both have the 2-8-0 wheel arrangement suited to the pulling of heavy trains.

Oakworth village Morris dancers perform their rousing act at Keighley station.

Above left: Here at Wansford on the Nene Valley Railway, former LNER D49 4-4-0 No. 62712 *Morayshire* has arrived from Scotland to make a rare appearance in the south.

Above right: Also at the Nene Valley Railway, classic Gresley and Thompson engines appear side by side. V2 2-6-2 No. 60800 *Green Arrow* prepares to take a service away from Wansford, while B1 class No. 1306 (BR 61306) *Mayflower* stands by.

Left: An auto-train with the engine in the middle of a two-carriage set prepares to move out of Berwyn station on the Llangollen Railway.

Steam railmotor No. 93 spends a weekend making excursions down the Brentford branch line from Southall. Here the unit passes under the now defunct footbridge outside Southall station, which was always a feature of any past photograph and a haven for trainspotters in the 1950s.

GWR diesel railcar No. 22 stands by to go into service at Didcot.

SOUTHERN CELEBRATES

Resplendent in full Edwardian 'Golden Age' liveries, H class 4-4-0 1905 SECR No. 263 (BR 31263) of the South Eastern & Chatham Railway (SE&CR) pulls No. 592 (31592) 1902 C class out of Horsted Keynes on the Bluebell Railway with a northbound train.

The celebrated SE&CR pairing head northwards towards the extended line to East Grinstead.

The only surviving example of the Urie-designed N15 London and South Western Railway (LSWR) class. The design was taken further by Maunsell for the SR and renamed 'King Arthurs' – it was used for express passenger work. No. 30777 *Sir Lamiel* (built by NBL Co. Ltd, Scotland, in 1925) is seen here outside the Loughborough works.

Ex-SE&CR C class No. 592 (BR 31592). The sumptuous Edwardian livery belies a massive achievement by all concerned in getting this locomotive back to life again from a scrapped and discarded condition.

RACES TO THE NORTH

Competition between the former private railway companies was intense. For those operating between London and Scotland, it was an on-going race to establish supremacy over the rivals by achieving ever-decreasing journey times.

Before the coming of the railways the quickest way to get from London to Scotland was by sea, which took twenty-four hours. A horse-drawn mail coach would take forty-eight hours. The innovation of the steam locomotive, running fast on tracks, changed people's lives forever. It was exciting, romantic and inspirational, and the Victorians loved it.

The emergence of a new breed of locomotive, the Stirling 4-2-2 Single, helped usher in a new era of fast northern travel. The bigger boilers and better steaming ability of these new locos took the railways into a new era, which the company owners were keen to exploit. Higher speeds and stronger pulling ability led to that competitive element coming to the fore. It was a race to see how quickly they could get to Edinburgh, which led to the 'great races to the north'. The London to Edinburgh service, which ran on the East Coast main line, and the London to Edinburgh via Glasgow, which ran on the West Coast main line, became the Blue Ribands of the railway services. But that wasn't the end of it. It was soon decided that a better race between the two rival lines would be definitive if the end location was to Aberdeen, where, because of the nature of the lines, there could be an undisputed winner.

Stroudley 'Terrier' 0-6-0T No. 32678 (known as No. 8 Knowle) at Tenterden.

SE&CR P class No. 753 in original livery glows in the early sunlight – note the immaculately kept facilities at Tenterden. This was another class designed for suburban work in London. Being slightly stronger, they took over some of the work of the Terriers.

BR Standard No. 73082 *Camelot* 5MT 4-6-0 (built in 1955 at Derby) heads a timetabled service away from Horsted Keynes on the Bluebell Railway.

The Standard Class 5s were a 172-strong series of locomotives. With two outside cylinders, their design was based on the renowned LMS Black 5s. The class sometimes took names from the former King Arthur class after they were scrapped.

The races to Edinburgh, on the other hand, were more time trials, of which there could be disputes over such things as running clearances and stops.

The West Coast route ran from London Euston via Crewe, Preston and Carlisle over the London and North Western Railway's track, before giving over to the Caledonian Railway. The East Coast route ran from London King's Cross on the Great Northern line via Doncaster, York and Newcastle to Edinburgh. Then it handed over to the North British Railway before it could head onwards across the Forth and Tay bridges to Kinnaber for the last stretch into Aberdeen.

Each route had the same London departure time: 10 a.m. for the Edinburgh run and 8.30 a.m. for Aberdeen. Both also had a night service to Aberdeen, leaving London at 8 p.m, and it was for this night service that the races became most competitive, as it was easier to get clearance priorities from signalmen and station staff who were getting caught up in the excitement of it all. Comfort stops were also reduced to a minimum, taking place only whilst there were engine and crew changes.

The key point on the Aberdeen run was Kinnaber Junction, 38 miles short of Aberdeen, where both lines converged onto one set of tracks. The first train to get signalled through the junction could not be beaten.

On the 23 August 1895 it was reported that the East Coast train won by fifteen minutes having arrived in Aberdeen at 4.40 a.m. after a trip of eight hours forty minutes (including stops). This was the first time the East Coast had won, which is surprising as the West Coast route is 16½ miles longer and has the severe gradients of Shap and Beatock.

This engine is No. 73082 *Camelot* and is shown here awaiting a programme of rebuilding to get a new ten-year running ticket. The BR Standard Class 5 is shown in a dismantled condition, but as shown in previous photographs it is now back in service.

The main long-term focus was, of course, to achieve ever-decreasing journey times for the 393-mile London to Edinburgh run. These achievements were extraordinary; the time was reduced from ten hours in 1869 to six hours in 1933, achieved by the *Coronation Scot* on a southbound west-coast run.

After a couple of years, the racing subsided, as it was taking its toll on man and machine. It was not until the new breeds of Gresley and Stanier locomotives came in that it was to start up again. Streamlining was all the rage in the 1930s, with the Coronations, Duchesses and A4s providing all the excitement.

The A1(3)-class engine 'Flying Scotsman' (which shares its name with the more famous train) made the first non-stop run in 1928. This was helped by the innovation of the walk-through tender, which allowed crews to be changed at the halfway point without the train stopping.

It's wartime at the Bluebell! With the evacuation of Dunkirk and the lead-up to D-Day, the SR was the busiest region through which almost everything military passed.

London, Brighton and South Coast Railway (LBSCR) Billington radial tank (radial axel below the cab) freshly overhauled and painted in the SR's 1920s version of light olive green. The engine was formerly called 'Birch Grove'.

The only example of the former 'Halt' sign still to be seen in use is on the approach road to Horsted Keynes station.

Another great southern preserved railway is at the Swanage Railway where Battle of Britain-class *Manston* passes the turntable on a home run into Swanage station.

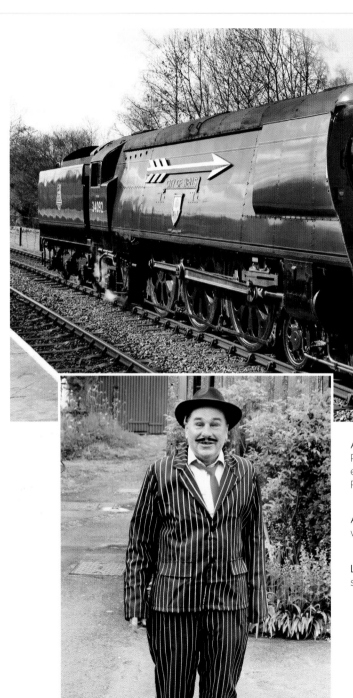

Above left: British Railways-liveried Bulleid 'West Country' Pacific 4-6-2 No. 34092 *City of Wells*. It was a Worth Valley engine that had an extended loan to the East Lancashire Railway, where it is seen here.

Above right: *City of Wells* backs out of Rawtenstall station where there are reminders of the Lancashire cotton industry.

Left: Children's entertainer Brillo is on duty at Kingscote station in Sussex.

GREAT NAMED TRAINS

The naming of trains really came to the fore between the wars when the era of great expresses demanded their own name and personality. The GWR started it in 1904 with the Cornish Riviera Express, the brainchild of Charles J. Churchward, its pioneering chief mechanical engineer. Cornwall had been well publicised as being the English riviera, where palm trees and exotic plants could be seen – even on stations.

The named trains became flagships of the different railway companies, always leaving from the same prominent platform at a precise time in the morning, seen off by a bowler-hatted stationmaster. The educated tones of the station announcer would never fail to mention the name several times in the lead up to the departure. The best staff would find their way to being allocated positions to which they proudly gave a superior service. The whole thing had style – each railway company offering its best and wanting everyone to know it. Unlike today, of course, where there is little room for character or style.

The daddy of all named trains is the *Flying Scotsman*, that great train which left London's King's Cross and Edinburgh's Waverley stations simultaneously at 10 a.m. every day. From 1928, when the A1 class of engine was available, the service became non-stop. At the time the journey would have taken seven and a half hours, across some 393 miles. A little later, when Sir Nigel Gresley introduced the classic A4s, a tender with a corridor was added so that the crew

Far left: Entertainment for everyone.

Left: The signal cabin at Horsted Keynes is right on the station, which enables the signalman to show visitors around (when there are no trains about, that is).

The sacrafices of RAF personnnel receive tributes at Horstead Keynes.

Right: An engine built in Brighton: Fairburn tank No. 42085 makes a visit to the Bluebell Railway and attracts plenty of attention.

could be changed mid run. The crew change was vital: a fireman would have to shift 5 tons of coal onto the fire over the complete journey. Quite the feat.

Through the war years and into the times of nationalisation there was little glamour surrounding the railways. It took the coronation of Queen Elizabeth to liven things up a bit when the Eastern Region of British Railways introduced a new non-stop train from London to Edinburgh and called it the *Elizabethan*. This was a summer-only service and became the longest non-stop service in Britain. Again the A4s fronted the trains, with their corridor tenders and quick crew change along the way.

The Cornish Riviera Express left from Platform 1 at the Brunel masterpiece that was Paddington station, London, with the first stop being Plymouth some 245 miles away. There were no change of crews for this one, so the fireman had a solid four hours' work emptying 4 tons of coal out of the tender and onto the fire – an all-time great railway achievement, which went largely without recognition. The engine and crew changed at Plymouth, as a King-class locomotive was too heavy to cross the Taymar bridge into Cornwall. The train went onto its final destination of Penzance, stopping several times on the way.

Also running from Paddington station in the 12 p.m. departure slot was the Torbay Express, an ideal train to be pulled by the GWR Castle-class locomotive. This train had a three-hour non-stop run to Exeter before going onto Torquay and Paignton. From there it would take the single-tracked branch line to finish at Kingswear (a section of line that is now the preserved Dartmouth Steam

Railway, which invites regular summer steam tours in memory of the Torbay Express).

The Pines Express was an unusually named train, for it was mostly noted for its runs on the Somerset & Dorset Joint Railway (S&DJR) – the famed cross-country route from Bath to Bournemouth over hilly terrain, but it actually set out from the industrial north to run via Manchester, Sheffield and Nottingham before arriving at Bath Green Park. At this location the engine would be changed before the train set off again in a reverse direction heading for Bournemouth West. The final outing for this train on the S&DJR was on the 8 September 1962 where it was hauled by 9F No. 90020 *Evening Star*, the last steam engine to be built by British Railways.

For a short time after the S&DJR closed, the Pines Express ran southwards via Oxford to Southampton, but after a few variations it was wound up in 1967.

The Master Cutler originated in 1947 to provide an express service between London Marylebone and the steel city of Sheffield. The train had a convoluted history, changing running patterns several times. It ran from different stations at both ends at various times, and is still running today. The preservation movement also likes to remind us of this icon with special outings, especially the Great Central Railway (GCR), on whose lines the original train ran.

The SR got well into the act by providing appropriately named trains to Brighton and Bournemouth – the former being the Brighton

Left: Electric sets were always the mainstay of the Southern Region. Here one is coming down the main line on the Great Central Railway. The four-car Southern Region Mk 1 4-BIG set propelled by the Class 33 (D6535) diesel at the rear, heading towards Quorn.

Below left and right: The prototype of Maunsell's Lord Nelson class No. 850 *Lord Nelson* at the Ropley shed on the Watercress line.

Belle, which came into operation as soon as the line was electrified in 1933.

Three five-car electric Pullman sets were manufactured with two sets in operation – one being kept as a spare. The umber and cream Pullman sets ran non-stop from London's Victoria station, taking just fifty-nine minutes to complete the 51 miles to Brighton, during which it reached a top speed of 80mph. The trains contained lush fabrics with each carriage interior being designed by a top, fashionable London store, encapsulating the art deco style of the day. The train ran for nearly forty years, ending up in the unglamorous times of 1972 where it was turned out in a mundane British Railways livery of grey and blue.

The South West of England was well represented by named trains, including the Atlantic Coast Express, a train that often more than duplicated itself in the holiday season. Leaving London's Waterloo station it would be fronted by a 'Lord Nelson' before the war, giving way to a Bulleid Merchant Navy-class locomotive afterwards. As the name implies, the train's final destination would be the northern towns of Cornwall and Devon, including Padstow, Bude and Ilfracombe. After Exeter the train would divide and sub-divide to serve a number of destinations in those westerly counties, often taken on by a Bulleid Light Pacific where that engine might be seen pulling just one carriage.

Left: The Lord Nelson-class engine has the Bulleid-period livery of malachite green; A4 LNER 'Pacific' *Bittern* is in the background.

Opposite: Metropolitan Railway 0-4-4 'E' No. 1 stands beside the watering point at Ropley station.

It was no surprise, then, that these tiny branch lines fell victim to Beeching's axe, leaving many areas devoid of a rail service.

In the same way that the *Flying Scotsman* ran the East Coast route to Scotland, the opposition provided the Royal Scot on the West Coast line. This train left London's Euston station daily at 10 a.m., as did a southbound train from Glasgow. For both lines, the heaviest and fastest locomotives were used, but with the distance being over 400 miles and the time taken just under eight hours, man nor machine could possibly hope to do it non-stop. In the 1920s the LMS developed the Royal Scot class of locomotive, but this had a power classification of just 7P, meaning that the engine and crew had to be changed at Carlisle. The LMS had to compete, so their chief mechanical engineer, Sir William Stanier, developed a monster: the Princess Royal class, the largest locomotive developed in Britain at the time. With a massive 45sq. ft firebox, this engine would have no trouble in making the trip – and breaking a few records along the way (such as Euston to Glasgow in five hours fifty-three minutes).

To follow the Princess Royals, Stanier also introduced the Coronation Scot class in 1935. These also were to target the West Coast route and had the streamline casing to rival the classic A4s of the LNER, later to be known as 'Duchesses'. The rivalry continued up to the outbreak of war, whereby everything became mundane. After the conflicts, the whole system was in such a sorry state that there was little interest in being competitive. At this time all the streamline casings were removed from the Duchesses.

All told there were over 200 named trains over the years distributed throughout the various regions. The GWR had the Bristolian and fast-running Cheltenham Flyer, the LNER had the Scarborough Flyer and Yorkshire

Pullmans, while the LMS came in with the Caledonian and Mid-day Scot. And no one could forget the Devon Belle of the SR, with its observation car to the rear.

A good many names have been kept alive by the modern main line railway companies, but it is particularly satisfying to see a titled train run by preservationists on the main network.

The Golden Arrow mustn't be forgotten – the pride of the SR and Southern Region (after nationalisation) – a luxury boat train with a destination of Paris. From London Victoria the train ran to Dover where the passengers were put on a ferry, and at Calais a similarly prestigious French train took them on to the Gare du Nord. Before the war it was the Lord Nelsons that fronted the train – until the coming of Bulleid's Merchant Navy class, that is. The insignia on the front is as impressive now as it was then: the gold arrow passing through the named disc with the flags of the two nations below, not forgetting the big arrows on the side. The service was electrified in 1961 but finished in 1972; however, the six-hour journey time was too much for it all to continue.

3

WEST COUNTRY SPECIALS

West Somerset action at Castle Hill: Black 5 No. 45231 *The Sherwood Forester* pilots rebuilt West Country class No. 34036 *Braunton* out of Williton and approaches the A358 before heading to Bishops Lydeard.

A linesider's dream: *The Sherwood Forester* and *Braunton* prepare to cross the A358 road bridge.

Above: The Torbay Express ran daily from London to the West Country, the final destination being Kingswear, which is where this train is approaching. The extension from Paignton to Dartmouth is now the highly successful Dartmouth Steam Railway.

Top right: No. 34046 *Braunton* is held at a red light to make way for the local steam train, headed by GWR tank engine No. 4277 2-8-0T *Hercules*. The main line express is headed by the 'West Country' rebuild. The former SR engine is standing in for a GWR Castle-class loco on this occasion.

Right: Recreating a scene from the past, the Torbay Express has arrived at its destination.

No. 34046 4-6-2 *Braunton* runs around the train and heads for Thurston to turn before taking the train back to Bristol. Meanwhile some of the passengers may enjoy a ferry crossing to Dartmouth.

Left: The steamer *Kingswear Castle* as seen from the ferry crossing of the River Dart.

Below left: GWR 57xx 0-6-0PT L.92 (GWR 75786) runs through Buckfastleigh on the South Devon Railway. No. L92 was part of a batch sold to London Transport between 1956 and 1963.

Below right: A view from the engine shed area of Bridgnorth station, a site which can accommodate up to forty engines. Western Region Pannier tank 0-6-0PT No. 1501 (1949) is in steam as a standby for a Severn Valley Railway Gala. In the foreground is Class 1400 tank No. 1450.

THE ORIGINAL GREAT CENTRAL

The original GCR was founded in 1899 by Edward W. Watkin, who died five years before its opening. Watkin was a visionary whose aim was to extend his northern railway network southwards with his own line into London, and then on to the Continent via a channel tunnel.

In creating the GCR, Watkin was competing with some major opposition: the Midland and Great Northern Railway and the North Western Railway. In order to compete successfully, and add another terminus to the line of stations above the royal boundary of Marylebone and Euston Roads in London (no railway was allowed to penetrate this boundary above ground), Watkin had to offer something extra. His solution was to offer a level of style and comfort that had never before been seen on the railways,

together with a completely new innovation – a buffet annex added to each carriage.

Following Watkin's death, another trendsetter was required to take the lead. Sam Fay became general manager of the GCR in 1902, bringing in new innovations and running the railway with great efficiency. GCR trains won a reputation for their speed, comfort and punctuality, and Fay was knighted for his services in 1912.

Watkin had earlier secured his railway's future by getting his foot in the door of other railways, with a seat on the board of the Metropolitan and South Eastern Railways, and the chairmanship of the Manchester, Sheffield & Lincolnshire Railway (MSLR). The MSLR crossed the whole of the north of England and came south as far as

The most powerful of the Panniers is the tapered boiler 9400 class. Introduced by Hawksworth, No. 9466 is seen here at Loughborough.

Wartime recreated on the forecourt of Kidderminster station. Note the black and white painted bollard, very useful in a blackout.

Annesley, near Nottingham. Watkin put a bill before Parliament to lay a new line from Annesley Road to Quainton Road in Buckinghamshire, conveniently at the end of the Metropolitan Railway. All that was left was to add in the short stretch at the London end, from Canfield Place to the new terminus. When the new railway opened, on 9 March 1899, the whole of Watkin's northern railway network was joined to form part of the GCR company.

The coming of the new railway caused much excitement around North London, with speculators gambling on where the new London terminus might be sited. One Frank Crocker was convinced that the location was going to be at Maida Vale, and built a stylish hotel to service it, but he is reported to have jumped to his death after the railway passed it by. The building, however, now a pub called Crocker's Folly, is still in business.

The GCR continued to operate until, like all the railways, it was taken over by the government for war service in 1914. After that it suffered the fate of almost all the independent railway companies, under the Grouping Act of 1921. It then became part of the LNER, under its new superintendent, Nigel Gresley. Signage at the preserved GCR still carries the colour of LNER blue.

An engine we've seen before. Black 5 No. 45231 *The Sherwood Forester* lets off considerable steam in the early morning sun at Minehead station.

GWR Manor class No. 7828 *Odney Manor*, temporarily carrying the name *Norton Manor* to honour the Marine base near the West Somerset Railway.

No. 7828 *Norton Manor* has the road to commence the run to Minehead; at 23 miles, this is the longest heritage line in the country.

The engineman conscientiously checks and prepares his engine for his next run.

Display of posters at Bishops Lydeard.

Black 5 No. 45231 *The Sherwood Forester* heads into Minehead.

Somerset and Dorset (S&D) Joint Railway Fowler No. 88 takes it easy behind No. 9351 as the double-header arrives at Blue Anchor from Minehead. Built at the Robert Stephenson Darlington works in 1925, the 2-8-0 was ideal to tackle the gradients on the S&D.

LNER A4 *Bittern* shows her graceful lines in the evening sunlight after a day's work on the West Somerset Railway.

Originally designed for London work by Beattie. Three of these well tanks survived owing to them working for nearly seventy years in the Cornish china clay industry. Well tank No. 30587 waits at the water facility at Bodmin General station, whilst No. 30585 runs through at Ropley on the Mid-Hants Railway. No. 30586 is still around but awaiting final restoration.

4

WARTIME AT NORTH YORKS AND NORTH NORFOLK

Black 5 No. 45428 *Eric Treacy* has arrived at Pickering on the North York Moors Railway. The engine now runs around the train beneath the new overall roof of the station.

Above left and right: The only survivor of a class that worked the mineral trains around Consett. Q6 0-8-0 No. 63395 climbs the notorious 1 in 49 gradient into Goathland station – not too difficult in mid-summer with this hardy preserved engine.

Left: BR Standard Mogul class 4 2-6-0 No. 76079 performed so well on main line outings that it became known as the 'Pocket Rocket'. Now a very useful engine in North Yorkshire, where it performs well on the Whitby extension.

The three-cylinder loco *The Great Marquess* is on duty for a wartime event. The autumn colours start to show as the train takes the curve into Levisham. The station has been renamed Le Visham for the war event, as it now represents a German-occupied station in France.

Activity at Pickering's market place.

ENGINES FOR WAR

Once the Ministry of Supply took control of the railways at the outbreak of the Second World War they quickly started copying procedures learned from operations during the First World War. The Robinson GCR 2-8-0 engine had been the most suitable locomotive for the heavy freight trains required for war transportation back in the earlier conflict, so it was the LMS 8F 2-8-0 that was selected as the standard machine for the Second World War. The wheel arrangement (2-8-0) was, it was decided, the most suitable to develop for hauling heavy goods trains, which were desperately needed at home and which would also later be required to replace wrecked equipment in Europe.

Passengers soon became aware that they were the lowest priority in the world of transport by train, as long delays and overcrowding became the norm for the duration. More locomotives therefore had to be designed and the man chosen to design and oversee production was Robert Riddles – the same man who had been part of the LMS team in charge of producing the successful 8F.

He prioritised another 2-8-0 engine to supplement the 8F, which this time was a product of austerity. From around 1943, 935 of the WD 2-8-0s were produced, with the belief that they would only have a lifespan of two years; they couldn't have been more wrong. Of these machines, 733 passed into the hands of British Railways

Le Visham – also known as Levisham – is seen here as an occupied French village station.

All is now at peace at Levisham after a Messerschmit has been brought down.

when the railways were nationalised in 1948, most of which lasted right up until the end of steam in Britain in the 1960s. But when it came to making a list of which machines were to be preserved, these trains appeared so humdrum that none of them made the cut. The Worth Valley Railway tracked down one example in Sweden that had escaped the cutter's torch. It managed to get it back to Keighley, where it can now be seen regularly heading trains along the line.

Although the 2-8-0 8Fs and WD versions were successful engines, their axle loading put limitations on their use, which demanded that another locomotive should be introduced.

The Ministry of Supply wanted a powerful, heavy freight engine for use at home and abroad. It needed to be able to work over lightly laid track, or lines with weak bridges, and therefore needed to have an axle loading of not more than 13 tons. Riddles answer was a 2-10-0 locomotive, of which 150 were built, all at the Hyde Park Works, Glasgow. When this became operational it was the first ever 2-10-0 design to run in Britain, as most did service in Britain before being taken abroad to help with the Allied liberation of Europe.

Some Spartan fittings were used in the manufacture of all these engines, such as a tiny chimney and cast front wheels. These gave the engines their utility appearance; they were designed for speed of manufacture, and prefabricated parts were used wherever possible. The only drawback with a set of ten coupled driving wheels was the radius of a curve that the engine could take. To help overcome this

Pickering's market place is host to a popular wartime event.

Wartime events continue with a parade through Pickering's market place.

restriction, the centre wheels were made flangeless and the second and fourth driving wheels were shallow-flanged. Other innovations were incorporated, such as a wide firebox and rocking grate for easy removal of the fire remains.

The North Yorkshire Moors Railway has been running two examples of this class, both of which came back to Britain from Greece in 1984: No. 90775 was formerly WD601 named 'Sturdee', while WD 3672 'Dame Vera Lynn' saw war service in Egypt.

One other notable example in preservation is No. 600 'Gordon'. Originally a training locomotive on the Longmoor Military Railway, this is now in storage with the Severn Valley Railway.

R.A. Riddles later was to become chief mechanical engineer (or the equivalent) to the newly formed British Railways in 1948. He was instrumental in producing the British Standard types, one of which was the 9F class 2-10-0. It was obvious from the many similarities that the 9F design was based on his wartime version.

'C'mon lads, it's only a little hill!' The last of the parade struggles through Pickering's market place.

The only steam bus still working runs around Whitby in the summer.

The Riddles-designed WD 2-10-0 90775 powers up the grade out of Sheringham, passing the golf course. Built in 1943, the loco was part of a batch of 100 built by North British Locomotive Co., Glasgow, for war use. She ended up in Egypt.

Former LNER D49 4-4-0 No. 62712 *Morayshire* makes an appearance at North Norfolk. Arriving at Weybourne, the train passes the on-platform signal cabin.

Above left: BR Class 4 2-6-0 No. 76084 prepares to change platform and backs up to the level crossing at Sheringham. Beyond this is the main line, which recently received a new connection after having been cut off for four decades.

Above right: The former David Shepherd engine 92203 *Black Prince* 2-10-0 waits for the starter signal at Sheringham before taking a service onwards to Holt. The engine is now resident at North Norfolk.

Left: BR Class 4 2-6-0 No. 76084 in action: the train is about to cross the A149 on the way to Weybourne.

1940s action at the North Norfolk Railway.

WOMEN PLAY THEIR PART

It was during the First World War that women were first employed to take over from men on the railways; approximately 190,000 men from the privately owned railways had enlisted. These jobs had to be filled and so the ladies stepped up, transforming social history. Women were to join a workforce that hitherto had been an all-male enclave. There may well have been women working before as cleaners or in canteens or within the station buffets, but none had been part of the operations force.

After some reservations from the men at the top, women were quickly put to use performing the same roles as their male counterparts, many of whom required specialist training, although some of those positions may have had particularly short training periods, such as ticket inspectors, guards and porters. The men in charge were slow to adapt to this culture shock.

Other areas of the railway industry would also witness the female touch: munitions (in railway workshops) and locomotive cleaning. Before it may have been normal for women to be seen on stations wearing skirts, but in these other roles the dress code would be entirely different. Hence, for the first time women could be seen in slacks or dungarees – another shock for the men of the railway. But with so many men leaving the railway to go to war, the whole rail system would have been unworkable without the introduction of female workers.

1940s action at the North Norfolk Railway.

The 1940s celebrations have spread into the streets of Sheringham.

At the end of the hostilities, women were encouraged to vacate their positions and go back to domestic duties and raise families; however, this was met with resistance. Many women with children were recently widowed and the railways were able to offer jobs where there might have been free accommodation, such as crossing keepers.

Through the first part of the war, women were paid two-thirds less than their male counterparts, something that was to change in later years. They also helped in the more hazardous areas of the work, such as platelayers on the track and in signal maintenance. Some female voices could also be heard over the station loudspeakers, which didn't go down well with the railway authorities. Clerical, telephone and telegraph were also areas of work that were well suited to the new female force.

Up until 1918 women were not allowed to vote in parliamentary elections – a situation that was changed by the Suffragette movement and their peaceful and not-so-peaceful campaigning – but it was not until 1928 that all women had the equal voting rights of men, which was enthusiastically received by this new force of female railway workers.

We know that not all of the men who went away to war came back, but the authority still tried to implement an Act to restructure the railways as they were before. This was met with opposition from women, with one poll showing that 83 per cent of women had voted to keep their employment. In the end, however, most were relieved of their duty, but an important mark had been made. This generation of women had experienced a new type of work and were reluctant to go back to their largely domesticated lives, although many of the women who stayed on in railway employment were offered lesser roles – which were also less well paid.

Two decades later, with the outbreak of the Second World War, there was no surprise, then, when the same vacancies as before became available. Throughout the late 1940s and '50s, women gained even greater employment opportunities, but it was not until 1958 that legislation was passed that gave women equal opportunities and equal pay.

1940s action at the North Norfolk Railway.

These ladies are really enjoying their day out in North Norfolk.

5

STEAM ON THE MAIN LINE

Two greats come together at York station on the 4 May 2014. LNER 4-6-2 No. 60009 *Union of South Africa* has arrived ten minutes early at York, where the engines are to be changed. LMS Coronation class 8P 4-6-2 (BR) No. 46233 *Duchess of Sutherland* is there to take over, where passengers take the opportunity to join the platform snappers for this historic steam meeting.

No. 46233 *Duchess of Sutherland* waits with her train before heading south to Peterborough and King's Cross.

No. 46233 *Duchess of Sutherland* heads a rail tour from Crewe towards Carlisle. On the outward journey at 11.20 a.m. the train takes the world-famous Settle to Carlisle route. Here we see the Duchess crossing the Ribblehead viaduct, a structure which might have been demolished had those who cherish our past not campaigned successfully to save it.

The decision to preserve the Settle to Carlisle line is vindicated by this twenty-one-wagon freight train heading southbound, taking up the entire length of the Ribblehead viaduct.

No. 46233 *Duchess of Sutherland* sits at the head of her train at Preston station. This classic station is located on the West Coast route to Scotland, the running ground of the London Midland & Scottish Railway, of which the *Duchess* was a prime user.

ENGINES AFTER THE WAR

With the state the country found itself in after the hostilities of the Second World War, it was certain that another decade would be needed for a sense of stability to return. However, there was to be even more austerity throughout this period, as the country was in debt and the whole railway system was worn out. Only essential repairs and renewal had been done during the war years; by 1946 it was time to count the costs.

Nationalisation had been considered after the First World War but was eventually passed up; this time, however, the descision would be different.

Where did this leave the stock of locomotives? They too were in a sorry state; the need for strip-down maintenance was immense, and many pre-grouping locos that had been kept on for the war would be scrapped as soon as new replacements could be brought along.

With nationalisation planned for 1 January 1948, the 'Big Four' private railway companies had their final chief mechanical engineers in place: Ivatt for the LMS; Peppercorn for the LNER; Hawksworth for the GWR and Bulleid for the SR. All these rail companies had been given back their control by the government after wartime, although the condition of their stock was nothing like that which had been handed over in 1939.

Hogwarts Express at Paddington: Harry Potter fans have the chance to take a magical steam-train ride from London's Paddington station. The Hogwarts Express is hauled by the GWR locomotive No. 5972 *Olton Hall*. Enthusiasts pose with the engine before boarding the train at Isambard Kingdom Brunel's magnificent station.

Definitely the right place for a Harry Potter fan. Departure is eagerly awaited at Paddington station.

The government handover came with a directive: build more engines, and quickly. With nationalisation looming there was little desire to design new types. Bulleid developed his Merchant Navy style with two new classes of Light Pacific, West Country and Battle of Britain classes. Hawksworth introduced the Counties, the structure of which was based on the wartime production of 8Fs, of which Swindon had produced a batch. He also built more of the Hall class, but upgraded these and called them Modified Halls. Other classic GWR types had been continued, such as the Castles and Manors, taking advantage of wartime developments to get better steaming from sub-standard coal. However, progress towards the oil-firing of engines had to be stopped when it was found that the country could not afford to pay for foreign oil. Other nations, which

had had their rail networks virtually wiped out, replanned from scratch, all leaning towards electric or diesel traction. Britain, whose railway was still intact, although run down, had to continue with steam to take advantage of the coal, which could be mined from Britain's own substructure.

Peppercorn, who had taken over from Edward Thompson in 1945, continued with his predecessor's plans, which in themselves were developments of Gresley designs and ideas. The Thompson B1 was very successful, with orders being placed for further numbers and a plan to develop Gresley's K4 class (of which an example, *The Great Marquess*, has been preserved) into a new 2-cylinder type was progressed. These became the K1s, of which seventy-one were finally released in 1949. All had the 'utility front', which indicates how

The 4-6-0 LMS Fowler/Stanier rebuilt Royal Scot class (BR) No. 46115 *Scots Guardsman* takes a rail tour for a steam-experience journey between Carnforth, Carlisle, Settle, Blackburn and Farington Junction in March 2014. Here the train rushes through the sleepy station of Long Preston.

Conveniently for the photographers, No. 46115 *Scots Guardsman* takes a water stop in the bay platform at Hellifield station.

Standing at Fort William station is LNER K1 class 2-6-0 No. 62005 (built 1949), which alternates with two Black 5s for the West Highland line steam motive power. The classification – 6MT – indicates the engine will be well suited to the sharp inclines it encounters on the line.

austerity was still the driver of policy. (No. 62005, a preserved K-1, regularly hauls the Jacobite train on the West Highland line.)

Peppercorn also designed the A2s, as there was now a shortage of express passenger engines (wartime had been all about building engines for heavy-goods haulage). Some of these designs had been attempts to redesign Gresley types P2 and V2, but with limited success and some ungainly looks. They did, however, get it right with the later version, one of which, No. 60532 *Blue Peter*, is preserved.

Forty-nine Peppercorn A1s came off the production line in the late 1940s, none of which were preserved for posterity. This gross oversight was put right by a group of dedicated enthusiasts who took it upon themselves to produce a new-build example. We now

have No. 60163 *Tornado* running on our lines, rolled out in 2008 to become one of the preservation treasures.

We should not forget the faithful saddle tanks produced in vast numbers during wartime by Hunslet and other manufacturers. Never has an engine pulled above its weight like these. Production continued after the war for a couple of years, and most of the preserved railways have one or two of these faithful engines to help stabilise their stock.

Ivatt of the LMS endeavoured to produce some new classes and came up trumps by looking abroad for ideas, producing the 2-6-0 tender engine and 2-6-2 tank version. These 2MT engines got named 'Mickey Mouse' by the enginemen of the time, due to their look of fragility, but they turned out to be great performers, which led to Robert Riddles continuing the style with his British Standard classes in the 1950s.

A train crosses the curving twenty-one-arch Glenfinnan viaduct, built in 1898 by 'Concrete Bob' McAlpine, founder of the famous construction company. This was a groundbreaking construction technique using unreinforced concrete to carry the single-track railway.

K1 2-6-0 No. 62005 sits at the buffers at Mallaig after a 40-mile journey. Passengers will alight for a couple of hours to browse around this little harbour village, the principle fishing port of western Scotland. Note the 'utility' front to the K1 where the running plate does not connect with the buffer beam plate.

Above left: The classic Cubitt front to King's Cross station, where the design shows the function of the building.

Above right: A visitor to the eastern region, West Country class No. 34046 *Braunton*, is standing in for A4 *Bittern* for a rail-tour trip to York and back. The classic double-barrel vaulted roof of the train shed contrasts with the single-span arch of the adjacent St Pancras station. On the far side is the suburban station.

Left: As previously seen at her home in Tyseley, GWR 4-6-0 No. 4965 *Rood Ashton Hall* is now out on the main line with the timetabled service, which operates throughout the summer. Here the train thunders through Stratford Parkway, with the return trip to Birmingham's Snow Hill station.

Above left and right: GWR Castle-class locomotive No. 5043 *Earl of Mount Edgcumbe* runs into Birmingham's Moor Street station with the final run of the day.

Left: The Gothic style reigns supreme on the cathedral-like frontage to Bristol Temple Meads station. The design reflects how rail travel at the time was perceived – marvellous and classic. The original station had outgrown Brunel's original structure, which still stands. The classic curved train shed and frontage was constructed between 1865 and 1876.

Above: Without any wheel slippage, rebuilt 'West Country' Light Pacific *Braunton* heads the train out of Bristol Temple Meads station, taking the famous Torbay Express train to its destination at Kingswear. It will need to negotiate the Wellington bank along the way, which will test the skills of the engine crew.

Left: The original service, which this train commemorates, was the daily 10.30 a.m. train from London Paddington to Kingswear. This 'West Country' engine No. 34046 *Braunton* heads off, passing below the steel-girder viaduct of Bristol's Bath Road.

MAIN LINE TRACK

On our railways we have a distance between the rails of 4ft 8½in, a gauge which has remained standard since the early nineteenth century. The tops of the rails soon lose their flatness as the traffic wheels running on them have a 9-degree camber to each wheel; the rims of the wheels of the rolling stock that run on the rails are raked outwards, which causes a similar camber to develop on the rails over time. The wheels also have inside flanges attached.

With the quality of the track we have, the wheels of the train that are made up of bogies (generally four-wheeled sets) can run at speed without the flanges coming into play. The bogies will run and take bends without the flanges having to touch the rails – on express lines there is a slight banking of the track to facilitate this. Of course, at slow speed, when switching tracks, the flanges are essential and sometimes a grinding whine can be heard if a slow train is taking a sharp bend.

Many of us will remember the clickety-clack as carriage wheels ran across joints between rails, but these can now be heard only on the preserved lines or in sidings. However, there was good reason for these gaps: it was so that in very hot weather the rails had a space to expand and, therefore, not buckle and deform. Traditionally rails would be joined together with bolted fishplates, but there are no gaps between modern main line tracks.

On the same day as the Torbay Express heads out of Bristol, another named train follows. This time LMS 5MT 4-6-0 No. 45407 *The Lancashire Fusilier* takes the 'Royal Duchy' on a steam tour to Cornwall and back.

Awaiting clearance to enter Kingswear terminus station. The 'West Country' rebuild No. 34046 *Braunton* has just arrived with a rail tour from Bristol. This sleepy corner of Devon once received the Torbay Express from London every day. Here that event has been reconstructed in a great heritage event.

The Whitby-bound train travels the length of the North York Moors Railway (NYMR) before arriving at the tracks of the main line at Grosmont. Here the train, headed by British Standard Class 4 4-6-0 No. 75029, is seen on the 'Levisham Straight', bringing the passengers along for the extended run.

The way this is now overcome is by pre-stretching the track lengths with hydraulic equipment at the time it is being laid. One length is stretched to the length it might attain if it were heated to something exceeding 27°C (in the UK). If the rails were not stretched and welded together stressing would take place as soon as the temperature rose significantly in hot weather. The former system of leaving gaps sufficed to cope with any expansion, as wear to the rails would happen firstly at the joint areas. Modern railways, though, need the gaps eliminated in order to provide smooth, continuous running to allow for heavier loads and higher speeds. The tracks become stressed either by the use upon them or through temperature changes, thereby undergoing tension or compression according to their rise or fall in temperature.

Once laid and welded, the track loses its original stable form. The modern rail is comprised of a high-grade steel alloy, which helps keep the liability of deformation to a minimum. With the lengths welded together, the rails could still expand and contract lengthways within the sockets of the holding shoes; this, however, is prevented by clips designed to hold the rail in place and minimise any movement. Ballast also needs to be in top order as this helps to stabilise the track from linear movement. Plastic or rubber pads are also inserted below the rails to discourage creep (the push on a rail that a train may make in the direction it is running) and to reduce the noise a train may make.

Steelworks these days can produce rails of very long lengths – well over 200m long. The rail lengths are loaded onto purpose-made transporting wagons that are as long as the rails. By design, rails will bend a surprising amount and main line railroads have rather low radius curves; therefore, these rail-carrying trains are able to travel on the rail network and get to their destination.

Continuous welded rail also acts as a signalling aid, whereby current can run along the track, and at points or interruptions a wire loop will be added to ensure continuity. Both rails act for differential directions of current flow. Therefore, if a wire is passed across the track to touch both rails at the same time, any green signal will turn to red; the axle of the train also performs the same function. Modern signalling generally keeps all signals at green, leaving the train itself to make its presence known and turn a green signal to red.

Effectively, a section of railway track is like an engineering I-beam, which spans from one sleeper to the next. It has to be able to take the load of the heaviest train; when that train is running at full speed, the stresses are much greater.

Rails with a flat bottom have proved the most successful. After stretching, the rails are laid when the train is moving to allow continuous unloading. Where a new rail butts up to one of the older type rails, an expansion link or 'breather' section can be added.

Track design, manufacture, laying and maintenance are separate sciences of their own, but all of those who work on rail manufacture uphold a proud railway tradition.

BR Standard Mogul Class 4 2-6-0 No. 76079 arrives at Grosmont bringing in a train off the Whitby extension (part of the national network).

On the approach to Whitby the train passes under the Larpool viaduct (now a cycle path and walkway) over the River Esk. This time the train is diesel hauled, with Class 25 D7628 *Sybilla* leading the way.

Above left: British Standard Class 4 No. 75029 *The Green Knight* stands at Whitby station. This engine was formerly owned by the artist David Shepherd, who bought it from British Railways and gave it its name.

Above right: Waiting with her train at Paddington station, the new-build A1 Peppercorn Pacific No. 60163 *Tornado* sports her third livery, after previously having been on show in apple green and the BR favourite of Brunswick green.

Left: After a quick dash on the electric train a photograph is snatched just in time at the next station – Royal Oak. Heading the 'Devon Belle' excursion the express powers forward past the location of the former Paddington turntable.

Above: It is now evening, where the Rambler has changed formation again to set off for what all the passengers have been waiting for. After a wait for clearance in the loop outside Bromsgrove station, the engines fire up to maximum for the attack on the Lickey Incline. The train passes into the station and has just 200m to get up to speed before climbing the 2-mile, 1 in 37 gradient unaided.

Above: The GWR Panniers No. 9600 and 7752 (L94) make a departure from their Tyseley base to take a 'Pannier Rambler' rail tour around the Midlands. Here the pairing is at full tilt coming through Kidderminster, where they give a whistle salute to the nearby Severn Valley Railway.

Right: GWR Castle class No. 5043 *Earl of Mount Edgcumbe* prepares to depart Birmingham Moor Street, with another running of the Shakespeare Express. Two round trips to Stratford make up the itinerary for these timetabled outings. The Castle and *Rood Ashton Hall* share the duties for this heritage steam service.

From a higher viewpoint *Earl of Mount Edgcumbe* heads out of Moor Street station in reverse for the second running of the day towards Stratford. This is part of Tyesley's Shakespeare Express summer timetabled excursions.

BIRMINGHAM
MOOR STREET

A fine December's day in 2015 sees Black 5 No. 45407 *The Lancashire Fusilier* enter West Hampstead station on the North London line with an excursion, which began the day at Southend. (West End Lane station was renamed West Hampstead in 1975.)

No. 5043 *Earl of Mount Edgcumbe* passes over the brick-arch viaduct in Birmingham.

The Black 5 No. 45407 loco was ex-works in 1937; the class has proved to be the supreme example of an all-purpose locomotive.

Heading for Salisbury, the above train passes a backdrop of a housing development at West Hampstead station. Class 47 Diesel No. 47580 *County of Essex* is providing some help (when needed) at the rear.

6

AROUND THE COUNTRY

No. 6023 *King Edward II* 4-6-0 was built in Swindon in 1930, as part of a series totalling just thirty. Here, coming down the main line from Loughborough, one of the three survivors makes an appearance at the Great Central Railway. The engine was rebuilt from scrap condition in an operation that took nineteen years. The engine has the livery of 1949–51 BR blue.

The prototype King, and pride of the former GWR, is at the national Railway Museum. No. 6000 is resplendent and still carries the bell it was given on a trip to America in 1927. Built at Swindon the same year, the series totalled just thirty examples.

In the confines of Moorgate station *Sarah Siddons* waits at the head of the train to Hammersmith until its departure at 12.25 p.m. These electric units ran between Baker Street and the outer reaches of Metroland from 1923 until their withdrawal in 1962. A total of twenty were built by Metropolitan Vickers.

On the back is Metropolitan locomotive 0-4-4T No. 1 E class. A gaggle of photographers at the end of the platform hurry to get in a picture before an incoming train arrives on the near platform to cut off the view.

Wartime engine at the Great Central Railway. Built at Ashford in 1943, LMS 8F No. 48624 was solely intended for the war effort, but once back in normal use the engine carried on until 1965.

LMS Railway Class 5 (Black Five) 4-6-0 approaches Quorn & Woodhouse from the north. Built by Armstrong Whitworth in 1937, No. 45305 was withdrawn from active service in 1968. Designed by Stanier, 842 examples of this class were built between 1934 and 1951.

OH, MR BEECHING!

Britain's railways were in financial turmoil. The Second World War had placed the whole railway system in peril, without the staff or the will to maintain it as a thriving clean and efficient service. The country had maintained steam at a point when other nations had opted for the cleaner and more efficient options.

The British Transport Commission (BTC) thought it was time to take some radical action to try to bring the railways into the modern world. In 1955, a plan was devised and published stating that all steam-powered locomotives would be removed from the system as soon as possible. (At this time the recently commissioned British Standard twelve classes of locomotives were still being produced in large numbers.)

Britain had invented the steam engine and was the producer of the greatest rail network in the world, which ran the best selection of engines in the world. These had been produced by private industrialists, with routes and services often duplicated, but the public loved them and the history that surrounded the evolution of the system is now a thing of intrigue.

Through the 1940s and '50s, schoolboys loved to spot the engines and underline the numbers in their Ian Allen spotter's books. Older people loved the glamour of the streamlined era and still relished the sheer theatre that surrounded all movements of a steam locomotive. Nevertheless, the time had come to do something about the decline. The BTC sought drastic action by employing someone from outside

Together with No. 1638 at Tenterden is British Standard 4 tank engine No. 80072, on loan from the Llangollen Railway. The engine appeared in all the documentaries featuring the London Tilbury and Southend line out of Fenchurch Street station, London.

Engine 57xx 0-6-0PT (GWR 7715) No. L99 was part of a batch sold to London Transport between 1956 and 1963. On 16 June 1971 engine number L94 took a train through the tunnels of the Metropolitan Railway, thereby signalling an end to steam services in Britain.

Introduced by Hawksworth (the last chief mechanical engineer of the GWR), Western Region Pannier tank 0-6-0PT No. 1501 (1949) is catching the afternoon sun at Holt, the southern terminus of the North Norfolk Railway.

the railway industry, a businessman with business and industrial knowledge: Dr Richard Beeching.

Beeching was on the board of Imperial Chemical Industries (ICI), the giant pharmaceutical corporation. He was a practical man who knew how to make things work. With the railways in the red – over £100m and spiralling upwards – it was time to cut away the parts of the system with the greatest losses.

This was also the time of the expansion of road transport, with the family car coming into the reach of the general population. Road transport had one great advantage over the railways in that they could load a lorry at one destination and deliver it directly to the recipient's address. There was none of this loading from a truck onto a railway wagon and then having to do all the offloading at the other end, with a British Railways 'mechanical horse' having to complete the delivery. The motorway network was well under construction by this time and freight had been steadily leaving the railways for more than a decade.

Beeching's former salary of £24,000 was matched – a huge sum, since the average working man's dream at the time was to reach £1,000 a year – which sealed his new appointment.

With the railway deficit running completely out of control, Prime Minister Harold Macmillan added his weight to the situation by backing the restructuring. He directed that Earnest Marples MP would be the government authority overseeing Beeching's proposals, who would then recommend the appropriate Acts of Parliament. This did not bode well for detractors of Beeching; Marples was a road man, having been in the business of road construction and with a keen eye for his own gain.

Beeching set forth on his largely unhindered crusade to reduce the system, and didn't hold back. After studying the situation and authorising some low-key on-the-spot surveys, he recommended that over 200 branch lines should be closed, with the removal of some 2,000 stations and 5,000 miles of track. This was to devastate some areas, especially the Highlands of Scotland, Wales and Devon and Cornwall, and other counties found that their railway map now showed huge gaps. Any objections or campaigns by local people, horrified by their loss of public services, largely fell on deaf ears.

The Beeching Report was published in March 1963, much to the surprise of the whole country. There had been a veil of secrecy over the proceedings at the BTC, with press clamouring, in vain, to have information fed to them. But the BTC had spoken: the railway map was to be changed for good. Bus services were supposed to take up the slack in providing transport connections for the people who had lost their trains, but these turned out to be less than enthusiastically introduced. Marples pushed everything through parliament, to leave the way clear for closure notices to be issued.

In 1965 Beeching left the railways and picked up a peerage.

Making an appearance at North Norfolk is this real gem of railway history: the Riley car owned by Ivo Peters, the exceptional photographer who covered the Somerset & Dorset Railway.

Another look at 'Modified Hall' No. 6990 4-6-0 *Witherslack Hall*, as the fireman has changed the points for the engine to run around the train. It is seen here at Leicester North station.

By 1973 his closure plans were just about complete. However, no forethought had been given to the fact that some of the smaller branch lines might have been useful again once the motorcar had overwhelmed every town and city in the land. Towns such as Hawick on the Scottish border were stranded – the Waverley line was one of the most controversial losses.

By chance, the events of the 1960s did supply us with something much more interesting than modernisation: the preservationist. In reaction against the speed and covert way in which everything once loved about railways was lost, there were the thousands of young people who stepped in to save some of Britain's heritage. We now have a preserved railway system, which is the envy of the world.

A new golden age has been reached where everyone can wallow in nostalgia.

The railways never did pay their way in the years following Beeching's plan. They continued forward and became more efficient, but all semblance of romance had gone. Diesel locomotives ran without names, and carriages became draughty and less comfortable. Fares went up – and up. The motorcar has become a victim of its own success, overwhelming the road network. How useful would some of those old branch lines be now?

All in all the transition could have been done better, but we did reach the modern age and at least we have preservation.

A visitor to the Keighley & Worth Valley Railway, 41312 4MT 2-6-2T is from the Watercress line. Here the Ivatt-designed LMS tank engine takes away a set of empty stock from Keighley station – another engine with a 'utility' front.

At the same location as the last picture, LMS 43924 has the all-clear to bring out the next service, with L&NWR coal tank No. 1054 taking the lead.

WD 90733 2-10-0 catches the morning sun as she reverses towards Keighley station to take out the first service of the day.

K4-class LNER (1938) *The Great Marquess* takes the curve outside Keighley station. The three-cylinder 4-6-0 engine has all the power needed for the 1 in 58 climb to Ingrow. She has often been seen in action on the Scottish West Highland line.

An engine we've seen in various locations around Britain, this British Standard 9F No. 92214 2-10-0 is a descendant from Riddles' wartime design, with the same wheel arrangement. Here it is seen at Swanwick Junction, the main location of the Midland Railway Centre.

The same engine as the previous photograph, now in the guise of 92207 *Evening Star* at Leicester North.

Coronation-class Pacific *Duchess of Sutherland* No. 46233 enters Preston station from the north. The 'Royal Scot' headboard represents the legendary former timetabled passenger service from London's Euston station to Glasgow (I.N.).

Newly built by a team of enthusiasts, Peppercorn A1 No. 60163 *Tornado* represents a recreation of an unpreserved class. Here the engine departs London's Victoria station. The engine's livery depicts the early days of nationalisation when the former LNER colour of apple green was allowed to continue.

The engine running previously as *Dominion of New Zealand* is seen here at Ropley workshops on the Watercress line under its own identity: A4 class 4464 (BR 60019) *Bittern*. Standing alongside it is 5MT Black 5 No. 45379.

In the summer of 2014 Britain had the pleasure of getting all six surviving A4 locomotives back together. Here are three of them at the National Railway Museum in York. Resident loco No. 60022 *Mallard* is in the centre flanked by 60010 *Dominion of Canada*, nearest, and 60008 *Dwight D. Eisenhower* on the far side.

No. 60022 *Mallard* with 60010 *Dominion of Canada* alongside. Note the bell on No. 60010 – essential for across the Atlantic, where the engine resides.

Gresley teak-bodied carriages on view at the North York Moors Railway.

Engine 9F 2-10-0 No. 92214 now has a home at the Great Central Railway, where it is seen here powering down the main line stretch towards Quorn.

Black 5 No. 44932 is on a cathedrals express rail tour in the east of England. Approaching Witham from the north the train has pulled off the former Great Eastern main line for a scheduled stop.

No. 44932 takes a water stop at Witham, Essex.

Locomotives require a special licence to run on the main line. Extra safety equipment has to be installed, such as Automatic Warning Systems (AWS).

The former signal gantry at Scarborough became obsolete and was offered to the North York Moors Railway, which, as we can see here, was accepted and put to fine use. The engine coming off the national rail line to Whitby is Black 5 No. 45428 *Eric Treacy*.

After a lengthy heatwave, the risk of lineside fires has become too much. Here at York, Class 47 Diesel No. 47580 *County of Essex* has had to deputise for a steam run and prepares to head for Scarborough. This fine station took a direct bomb hit in the Baedeker raids of April 1942.

The rare paring of two engines with their roots in Victorian times: *City of Truro* coupled with *Earl of Berkeley* double-head a service into Ropley arriving from Alresford.

Left: GWR Dukedog *Earl of Berkeley* takes the lead position for a double-headed excursion along the Watercress line.

Below left: The *City of Truro* is seen here in glorious Edwardian livery as she rolls along to the shed at Torrington on the Gloucestershire Warwickshire Railway. The double-framed 4-4-0 design clearly shows the origins of this remarkable loco.

Below right: Shafts of early sunlight catch the *Flying Scotsman* and her train as she powers upgrade after a water stop at Connington. The occasion is the inaugural main line outing in February 2016 after a ten-year overhaul – next stop Peterborough on the King's Cross–York schedule. Note the overhead cables supported by wires rather than the heavier goalpost-type supports.

RAILWAY DIRECTORY

Steam Guide (Standard Gauge)

Alderney Railway
Alderney. C.I.
www.alderneyrailway.com
Tel: 01455 634373

Avon Valley Railway
Bitton Station. Nr Bristol. 6HD
www.avonvalleyrailway.org.uk
Tel: 01457 484950

Barrow Hill Roundhouse
Chesterfield. Derbyshire. S43 2PR
www.barrowhill.org.uk
Tel: 01246 472450

Barry Tourist Railway
www.barrytouristrailway.co.uk
Tel: 01446 748816

Battlefield Line Railway
Shackerstone Station. CV13 6NW
www.battlefield-line-railway.co.uk
Tel: 01827 880754

Bluebell Railway
Sheffield Park Station. TN22 3QL
Horsted Keynes Station. RH17 7BB
www.bluebell-railway.co.uk
Tel: 01825 720800

Bodmin and Wenford Railway
Bodmin General Station. PL31 1AQ
www.bodminrailway.co.uk
Tel: 01208 73666

Bo'ness and Kinneil Railway
Bo'ness Station. EH51 9AQ
www.bkrailway.co.uk
Tel: 01506 825855

Bowes Railway Centre
Gateshead. NE9 7QJ
www.newcastlegateshead.com
Tel: 01914 161847

Bressingham Steam Museum
Nr Diss. Norfolk. IP22 2AA
www.bressingham.co.uk
Tel: 01379 686900

Bristol Harbour Railway
Princes Wharf. BS1 4RN
www.bristolharbourrailway.co.uk
Tel: 01173 526600

Buckingham Railway Centre
Quainton Road Station. HP22 4BY
www.bucksrailcentre.org.uk
Tel: 01296 655720

Caledonian Railway
Brechin Station. DD97AF
www.caledonian-railway.com
Tel: 01356 622992

Chasewater Railway
Brownhills West. WS8 7NL
www.chasewaterrailway.co.uk
Tel: 01543 452623

Chinnor & Princes Risborough Railway
Chinnor Station. OX39 4ER
www.chinnorrailway.co.uk
Talking Timetable: 01844 353535

Cholsey & Wallingford Railway
Wallingford. OX10 9GQ
www.cholsey-wallingford-railway.com
Tel: 01491 835067

Churnet Valley Railway
Kingsley & Froghall Station. ST10 2HA
www.churnet-valley-railway.org.uk
Tel: 01538 750755

Colne Valley Railway
Castle Hedingham. CO9 3DZ
www.colnevalleyrailway.co.uk
Tel: 01787 461174

Darlington Railway Museum
Station Rd. Darlington. DL3 6ST
www.darlington.gov.uk
Tel: 01325 460532

Dartmoor Railway
Oakhampton. EX20 1EJ
www.dartmoorrailway.com
Tel: 01837 55164

Dartmouth Steam Railway & River Boat Company
Paignton. TQ4 6AF
www.dartmouthrailriver.co.uk
Tel: 01803 555872

Dean Forest Railway
Lydney. GL15 4ET
www.deanforestrailway.co.uk
Tel: 01594 845840

Derwentvalley Light Railway
Murton Park, York. YO19 5UF
www.dvlr.org.uk
Tel: 01904 489966

Didcot Railway Centre
Didcot. OX11 7NJ
www.didcotrailwaycentre.org.uk
Tel: 01235 817200

Downpatrick & Co. Down Railway
Downpatrick Station. N.I. BT30 6LZ
www.downrail.co.uk
Tel: 02844 612233

East Anglian Railway Museum
Chappel, Near Colchester. CO6 2DS
www.earm.co.uk
Tel: 01206 242524

East Kent Railway
Shepherdswell. CT15 7PD
www.eastkentrailway.co.uk
Tel: 01304 832042

East Somerset Railway
Cranmore Station. BA4 4QP
www.eastsomersetrailway.co.uk
Tel: 01749 880417

East Lancashire Railway
Bury Bolton Street Station. BL9 0EY
Rawtenstall Station. BB4 6DD
Ramsbottom Station. BL0 9AL
www.eastlancsrailway.org.uk
Tel: 01617 647790

Ecclesbourne Valley Railway
Wicksworth. DE4 4FB
www.e-v-r.com
Tel: 01629 823076

Elsecar Heritage Railway
Elscar Heritage Centre. S74 8HJ
www.elsecarrailway.co.uk
Tel: 01226 740203

Embsay & Bolton Abbey Steam Railway
Bolton Abbey Station, Skipton. BD23 6AF
www.embsayboltonabbeyrailway.org.uk
Tel: 01756 710614
Talking Timetable: 01756 795189

Epping Ongar Railway
Ongar Town. CM5 9AB
www.eorailway.co.uk
Tel: 01277 365200

Foxfield Steam Railway
Blythe Bridge. ST11 9BG
www.foxfieldrailway.co.uk
Tel: 01782 396210

Gloucestershire Warwickshire Railway
Toddington Station. GL54 5DT
Winchcombe Station. GL54 5LB
www.gwsr.com
Tel: 01242 621405

Great Central Railway
Loughborough Central Station. LE11 1RW
Quorn & Woodhouse. LE12 8AW
Leicester North. LE4 3BR
www.gcrailway.co.uk_
Tel: 01509 632323

Great Central – Nottingham
Ruddington. NG11 6JS
www.gcrn.co.uk
Tel: 0115 9405705

Gwili Railway
Carmarthen. SA33 6HT
www.gwili-railway.co.uk
Tel: 01267 230666

GWR (Steam Museum)
Swindon. SN2 2EY
www.steam-museum.org.uk
Tel: 01793 466646

Isle of Wight Steam Railway
Havenstreet. PO33 4DS
www.iwsteamrailway.co.uk
Tel: 01983 882204

Keighley & Worth Valley Railway
Haworth Station. BD22 8NJ
Keighley Station. BD21 4HP
www.kwvr.co.uk
Tel: 01535 645214

Kent & East Sussex Railway
Tenterden Station. TN30 6HE
www.kesr.org.uk
Tel: 01580 765155.
Talking Timetable: 01580 762943

Lakeside & Haverthwaite Railway
Haverthwaite Station. LA12 8AL
www.lakesiderailway.co.uk
Tel: 01539 531594

Lavender Line
Isfield Station. TN22 5XB
www.lavender-line.co.uk
Tel. 01825 750515

Lincolnshire Wolds Railway
Ludborough. DN36 5SH
www.lincolnshirewoldsrailway.co.uk
Tel: 01507 363881

Llangollen Railway
Llangollen Station. LL20 8SN
www.llangollen-railway.co.uk
Tel: 01978 860979

Mangapps Farm Railway Museum
Burnham-on-Crouch. CM0 8QG
www.mangapps.co.uk
Tel: 01621 784898

Middleton Railway
Hunslet. LS10 2JQ
www.middletonrailway.org.uk
Tel: 0845 680 1758

Mid-Norfolk Railway
Dereham Station. NR19 1DF
www.mnr.org.uk
Tel: 01362 851723

Mid-Hants Railway (Watercress Line)
Alresford Railway Station. SO24 9JG
Ropley Station. SO24 0BL
www.watercressline.co.uk
Tel: 01962 733810

Midland Railway Centre
Butterley Station. DE5 3QZ
Swanwick Junction.
www.midlandrailwaycentre.co.uk
Tel: 01773 570140

Mid-Suffolk Light Railway
Wetheringsett. IP14 5PW
www.mslr.org.uk
Tel: 01449 766899

National Railway Museum
Leeman Road.York. YO26 4XL
www.nrm.org.uk
Tel: 0844 815 3139

Nene Valley Railway
Wansford Station. PE8 6LR
www.nvr.org.uk
Tel: 01780 784444

Northampton & Lamport
RAILWAY
Pitsford & Brampton Station. NN6 8BA
www.nlr.org.uk
Tel: 01604 820327

Northamptonshire Ironside Railway Trust
Northampton. NN4 9UW
www.nirt.co.uk
Tel: 01604 702031

North Norfolk Railway
Sheringham Station. NR26 8RA
Holt Station. NR25 6AJ
www.nnrailway.co.uk
Tel: 01263 820800

North Tyneside Steam Railway
(Stephenson Railway Museum)
North Shields. NE29 8DX
www.twmuseums.org.uk
Tel: 0191 2007146

North York Moors Railway
Pickering. YO18 7AJ
Goathland. YO22 5NF
Grosmont. YO22 5QE
www.nymr.co.uk
Tel: 01751 472508

Pallot Steam, Motor & General Museum
Jersey. C.I.
www.pallotmuseum.co.uk
Tel: 01534 865307

Peak Rail
Matlock. DE4 3NA
www.peakrail.co.uk
01629 580381

Plym Valley Railway
Plympton. PL7 4NW
www.plymrail.co.uk
Tel: 07580 689380

Pontypool & Blaenavon Railway
Blaenavon. NP4 9ND
www.pontypool-and-blaenavon.co.uk
Tel: 01495 792263

Railway Preservation Society Of Ireland
www.steamtrainsireland.com
Tel: 028 9337 3968

Rutland Railway Museum
Cottesmore. LE15 7BX
www.rutnet.co.uk
Tel: 01572 813203

Severn Valley Railway
Bridgnorth. WV16 5DT
Bewdley. DY12 1BG
Kidderminster. DY10 1QX
www.svr.co.uk
Tel: 01299 403816

South Devon Railway
Buckfastleigh. TQ11 0DZ
www.southdevonrailway.co.uk
Tel: 0843 357 1420

Southall Railway Centre
Southall. UB2 4SE
www.gwrpg.co.uk
Tel: 0208 5741529

Spa Valley Railway
Tunbridge Wells. TN2 5QY
www.spavalleyrailway.co.uk
Tel: 01892 537715

Strathspey Railway
Aviemore. PH22 1PY
www.strathspeyrailway.co.uk
Tel: 01479 810725

Swanage Railway
Swanage. BH19 1HB
www.swanagerailway.co.uk
Tel: 01929 425800

Swindon & Crickslade Steam Railway
Blunsdon. Wilts.
www.swindon-crickslade-railway.org.uk
Tel: 01793 771615

Tanfield Railway
Gateshead. NE16 5ET
www.tanfield-railway.co.uk
Tel: 0845 463 4938

Telford Steam Railway
Horsehay. TF4 2NG
www.telfordsteamrailway.co.uk
Tel: 01952 503880

Tyseley Railway Centre
(Vintage Trains)
670 Warwick Road. Tyseley. B11 2HL
www.tyseleylocoworks.co.uk
Tel: 01217 084960

West Somerset Railway
Minehead Station. TA24 5BG
Williton Station. TA4 4RQ
Bishops Lydeard. TA4 3RU
www.westsomersetrailway.co.uk
Tel: 01643 704996

The History Press
The destination for history
www.thehistorypress.co.uk